MEATS AND PROTEINS

by Robin Nelson

Lerner Books · London · New York · Minneapolis

We need to eat many kinds of food to be healthy.

We need to eat foods in the **meat** and **protein** group.

We can eat meat, **poultry**, fish, **pulses**, eggs and nuts.

These foods give us protein and **fat**.

Protein helps build our bodies.

Fat gives us energy and keeps our bodies warm.

We need two **servings** of meat and protein each day.

We can eat a hamburger.

We can eat fish.

We can eat turkey.

We can eat prawns.

We can eat pulses.

We can eat eggs.

We can eat peanut butter.

We can eat nuts.

Meat and protein foods
keep me healthy.

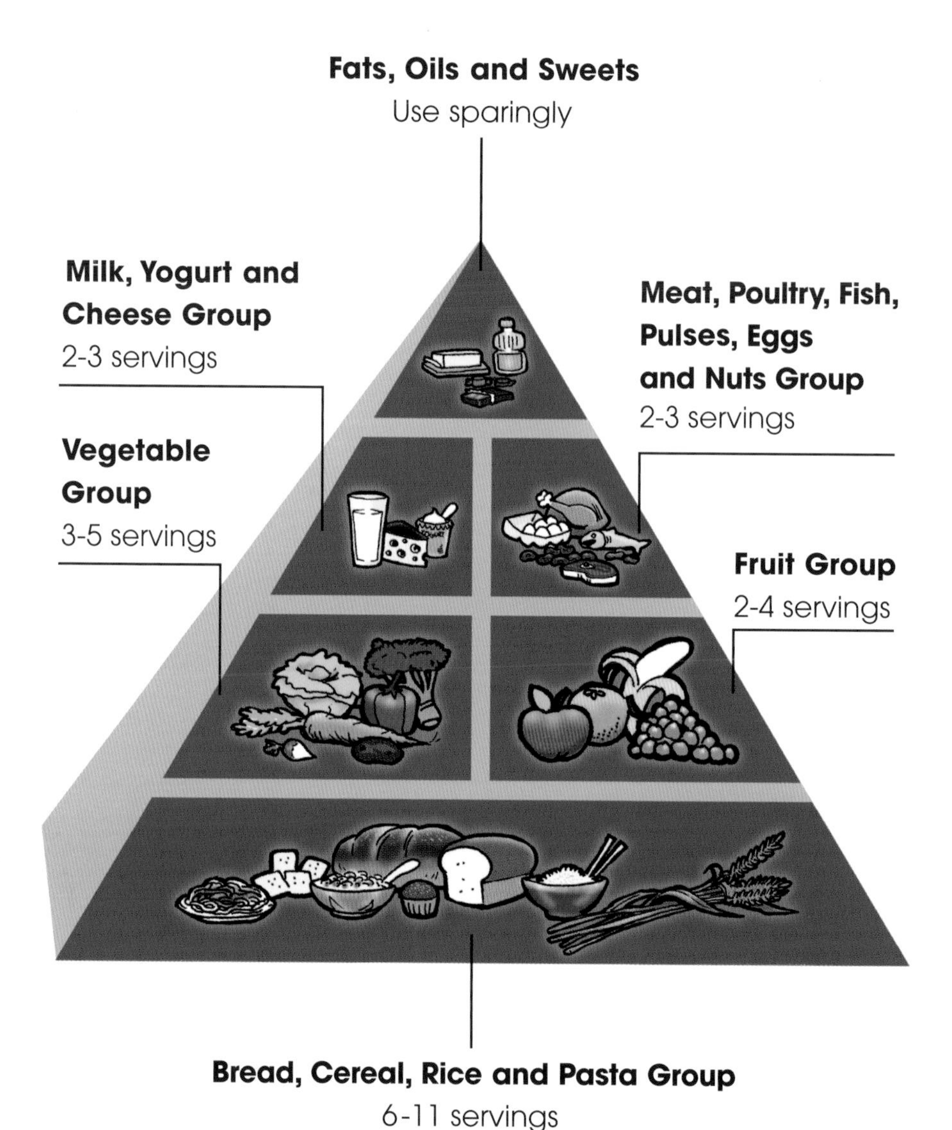
Fats, Oils and Sweets
Use sparingly
Milk, Yogurt and
Cheese Group
2-3 servings
Meat, Poultry, Fish,
Pulses, Eggs
and Nuts Group
2-3 servings
Vegetable
Group
3-5 servings
Fruit Group
2-4 servings
Bread, Cereal, Rice and Pasta Group
6-11 servings

Meat, Poultry, Fish, Pulses, Eggs and Nuts Group

The food pyramid shows us how many servings of different foods we should eat every day. The meat, poultry, fish, pulses, eggs and nuts group is on the third level of the food pyramid. You need 2–3 servings from this group every day. The foods in the meat, poultry, fish, pulses, eggs and nuts group are good for you because they have protein. Protein helps our bodies grow.

Meat and Protein Facts

British farmers produce enough lamb for more than 250 million Sunday roasts in a year.

The peanut is not a nut. It is in the pulses group and belongs to the pea family.

There are more chickens in the world than people.

Hens lay one egg every 24 to 26 hours.

Beef comes from cattle and gives our bodies iron, zinc, protein and B-vitamins. These nutrients are an important part of a healthy diet.

Pork is meat we get from pigs. Pigs give us pork chops, ham, bacon, spare-ribs and sausages.

Lamb is meat we get from sheep.

Glossary

fat – parts of food that give you energy

poultry – meat from birds like chickens, turkeys, ducks and geese

protein – parts of food that give us energy and help build bones, hair, muscles and skin

pulses – seeds and beans like lentils, chick peas and butter beans

servings – amounts of food

Index

This book was first published in the United States of America in 2003.

First published in the United Kingdom in 2008 by
Lerner Books,
Dalton House,
60 Windsor Avenue,
London SW19 2RR

Website address: www.lernerbooks.co.uk

This edition was updated and edited for UK publication by Discovery Books Ltd.,
Unit 3, 37 Watling Street, Leintwardine, Shropshire SY7 0LW.

Words in **bold** type are explained in the glossary on page 22.

British Library Cataloguing in Publication Data

Nelson, Robin, 1971-
Meats and proteins. - (First step non-fiction. Food groups)
1. Meat - Juvenile literature 2. Proteins in human
nutrition - Juvenile literature
I. Title
641.3'6

ISBN-13: 978 1 58013 391 3

The photographs in this books are reproduced through the courtesy of: © Todd Strand/ Independent Picture Service, front cover, pp 3, 4, 5, 13, 15, 16, 22 (top, second from bottom, bottom); © PhotoDisc/Getty Images, pp 2, 6, 8, 9, 11, 12, 22 (second from top, middle); © Royalty-Free/CORBIS, pp 7, 14; © USDA/Ken Hammond, p 10; © Rubberball Productions, p 17.

The illustration on page 18 is by Bill Hauser/Independent Picture Service.

Printed in China